# Microbes at Biomedical Interfaces 2020



Topical Conference at the 2020 AIChE Annual Meeting

Online
16 – 20 November 2020

ISBN: 978-1-7138-2290-5

**Printed from e-media with permission by:**

Curran Associates, Inc.
57 Morehouse Lane
Red Hook, NY 12571

**Some format issues inherent in the e-media version may also appear in this print version.**



AIChE
120 Wall Street, FL 23
New York, NY 10005-4020

Phone: (800) 242-4363
Fax: (203) 775-5177

www.aiche.org

**Additional copies of this publication are available from:**

Curran Associates, Inc.
57 Morehouse Lane
Red Hook, NY 12571 USA
Phone: 845-758-0400
Fax: 845-758-2633
Email: curran@proceedings.com
Web: www.proceedings.com

# TABLE OF CONTENTS

(99A) PREVENTING BACTERIAL COLONIZATION OF SURFACES WITH LAYER-BY-LAYER COATINGS OF FLUORINATED POLYPHOSPHAZENES AND ANTIBIOTICS ........................ 1
*Svetlana A. Sukhishvili, Victoria Albright, Alexander K. Andrianov*

(99B) TREATMENT OF GRAM-NEGATIVE BACTERIA USING STIMULI-RESPONSIVE NANOANTIBIOTICS ........................ 2
*Julius Edson, Young Jik Kwon*

(99C) NANOSUS (ULTRA FINE GRAINED STAINLESS-STEEL) FOR ORTHOPEDIC IMPLANTS ........................ 3
*Shaunak Kelkar, Bahram Saleh, Caterina Bartomeu Garcia, Fumie Yusa, Takafumi Komatsu, Thomas J. Webster*

(99D) B-LACTAMASE RESPONSIVE HYDROGELS FOR BACTERIA-TRIGGERED ANTIBACTERIAL TREATMENTS ........................ 5
*Dahlia Alkekhia, Chao Yu, Cassi Larose, Anita Shukla*

(99E) ZWITTERIONIC AND FLUORINATED AMPHIPHILIC POLYMER COATINGS WITH ENHANCED RESISTANCE TO BIOFILM FORMATION ........................ 6
*Trevor Donadt, Rong Yang*

(99G) INHIBITING BACTERIAL SURFACE COLONIZATION USING PEPTOID-LOADED POLYELECTROLYTE MICROGELS ........................ 7
*Wenhan Zhao, Xixi Xiao, Jennifer S. Lin, Annelise Barron, Matthew Libera*

(99H) FABRICATION OF OIL-INFUSED ANTI-BIOFOULING COATINGS ON THE SURFACES OF FLEXIBLE POLYMER TUBING ........................ 9
*Harshit Agarwal, Kayleigh E. Nyffeler, Timothy A. Hacker, Sean P. Palecek, Helen E. Blackwell, David M. Lynn*

(99I) MODIFYING PLASTIC SURFACES IN DAIRY MEMBRANES TO ACT AS ANTIMICROBIAL SURFACES ........................ 10
*Stephen Ritchie, Ryan M. Summers, William D. Baker*

(99J) INTERFACIAL PROTEIN SELF-ASSEMBLY: AN "IRONIC" APPROACH TOWARDS GENERATING ANTIFOULING COATINGS ........................ 11
*Tanner D. Fink, Peyman Delparastan, Joschka Bauer, Thomas Scheibel, Phillip Messersmith, Runye Zha*

(193A) CORRELATING STAPHYLOCOCCUS EPIDERMIDIS BIOFILM PH GRADIENTS WITH LOCAL BIOFILM MECHANICAL PROPERTIES TO REVEAL CONTRIBUTIONS OF PH TO BIOFILM DISASSEMBLY AND DISPERSION ........................ 12
*Patryck Michalik, Elizabeth J. Stewart*

(193B) THERMALLY TRIGGERED BIOFILM DISPERSION ........................ 13
*Haydar Aljaafari, Parham Parnian, Hannah Chicchelly, Eric Nuxoll*

(193D) BIOINTERFACIAL DESIGN TO MANIPULATE SIDEROPHORE PRODUCTION BY PSEUDOMONAS AERUGINOSA WITHOUT SOLUBLE CHEMICAL INTERVENTION ........................ 14
*Trevor Donadt, Rong Yang*

(193E) CONTROLLING THE ADHESION OF STAPHYLOCOCCUS AUREUS TO POLYDIMETHYLSILOXANE BIOMATERIALS ............ 15
*Irene S. Kurtz, Abraham J. Waldman, Jessica D. Schiffman*

(193H) MASS-TRANSFER RELATIONS FOR FOR NANOPARTICLE TRANSPORT ACROSS HUMAN-BIOFILM INTERFACE ............ 16
*Joshua Prince, A-Andrew Jones*

(245A) SALIVARY MUCIN GLYCANS SUPPRESS QUORUM SENSING AND ASSOCIATED VIRULENCE TRAITS IN STREPTOCOCCUS MUTANS ............ 17
*Caroline Werlang, Wesley Chen, Kazuhiro Aoki, Michael Tiemeyer, Katharina Ribbeck*

(245B) CHARACTERIZATION OF INTERFACIAL FILMS OF MUCOID AND NONMUCOID PSEUDOMONAS AERUGINOSA ISOLATES ............ 18
*Sricharani Balmuri, Vienvilay Phandanouvong-Lozano, Tagbo H. R Niepa*

(245D) NANOTOPOGRAPHICAL AMPLIFICATION OF BACTERIA-MATERIAL REPULSIVE INTERACTIONS FOR MITIGATING BACTERIAL ATTACHMENT AND BIOFILM FORMATION ............ 19
*Yifan Cheng, Guoping Feng, Diana A. Borca-Tasciuc, Randy W. Worobo, Carmen I. Moraru*

(245E) INTERACTION OF CYANOBACTERIA WITH NANOMETER AND MICRON SIZED POLYSTYRENE PARTICLES IN MARINE AND FRESH WATER ............ 20
*Tania Oliveira, Irene Andreu Blanco, Mary C. Machado, Anubhav Tripathi, Arijit Bose*

(245F) USING FREELY SUSPENDED BIOFILMS TO STUDY INTERACTIONS AMONG BACTERIAL CELLS: MECHANICAL AND TRANSPORT PROPERTIES ............ 21
*Priya Chittur, Hanwei Liu, David Tirrell, Julia A. Kornfield*

(245G) ZWITTERION-CONTAINING COPOLYMERS REDUCE BACTERIAL ADHESION IN DENTAL ENVIRONMENTS ............ 22
*Irene S. Kurtz, Ashlin Sathyan, Todd Emrick, Jessica D. Schiffman*

(245I) NOVEL COATING MATERIAL FOR CONTROLLING CORROSION AND BIOFOULING IN WATER SUPPLY SYSTEM ............ 23
*Qiong Wu, Wei Han, King Lun Yeung*

(330A) DIRECT IMAGING METHODS SHOW LIMITATIONS OF TRADITIONAL BIOFILM STAINING TECHNIQUES ............ 24
*Abner Bogan*

(330B) MULTIFUNCTIONAL NANOCULTURE SYSTEM FOR STUDYING MICROBIAL DYNAMICS ............ 25
*Huda Usman, Aissatou Bah, Tagbo H. R Niepa*

(330C) UNDERSTANDING THE ROLE OF PROTEIN DEPOSITION IN MICROBIAL INTERACTIONS WITH CATHETERS USING LIQUID-INFUSED SURFACES ............ 26
*Junie Fong, Marissa Andersen, Ana L. Flores-Mireles, Caitlin Howell*

(330D) REVEALING GENETIC FUNCTIONS DURING BACTERIAL ADHESION AND BIOFILM FORMATION USING BARCODE TRANSPOSON MUTANT LIBRARY ............ 27
*Fangchao Song, Sylvia Illouz, Adam M. Deutschbauer, Adam P. Arkin*

(330E) ELUCIDATING A NEW EXTRACELLULAR FUNCTION OF DEGP INHIBITING BIOFILM FORMATION OF ENTEROHEMORRHAGIC ESCHERICHIA COLI O157:H7 ............ 28
*Kuili Fang, Seok Hoon Hong*

(330G) ANTIFOULING TOPOGRAPHIES TO COMBAT BIOFILM-ASSOCIATED ANTIBIOTIC TOLERANCE ..... 29
*Huan Gu, Sang Won Lee, Dacheng Ren*

(330H) COMPUTATIONAL MODELING AND EXPERIMENTAL CHARACTERIZATION OF CATHODIC VOLTAGE CONTROLLED ELECTRICAL STIMULATION (CVCES) FOR ERADICATION OF BIOFILMS ..... 31
*Priyanshu Vishnoi, Caelen Clark, Mark Ehrensberger, Mark T. Swihart*

(330I) INTERFACEOF ELECTROGENIC BACTERIA AND REDUCED GRAPHENEOXIDE: ENERGETICS AND ELECTRON TRANSPORT ..... 32
*Sheldon Cotts, Bijentimala Keisham, Vikas Berry*

(93B) FINDING A NICHE – MICROBIAL COLONIZATION OF HOST MUCOSAL BARRIERS ..... 33
*Katharina Ribbeck*

(93A) LIFE IN A TIGHT SPOT: TRANSPORT AND COLLECTIVE BEHAVIOR OF BACTERIA IN HETEROGENEOUS MEDIA ..... 34
*Sujit S. Datta*

(93D) THE PSEUDOMONAS AERUGINOSA BIOFILM MATRIX ..... 35
*Matthew Parsek*

(390A) SPATIAL LOCALIZATION AND TREATMENT OF INDIVIDUAL CELLS IN MULTISPECIES BIOFILMS ..... 36
*Nuno Azevedo*

(390B) BIOLOGICAL INSPIRATION FOR ANTI-BIOFILM STRATEGIES ..... 37
*Allon I. Hochbaum*

(390C) BIOMIMETIC DRUGS: IN VITRO, IN VIVO & MECHANISTIC STUDIES OF 5-12MER ANTIMICROBIAL & ANTIVIRAL PEPTOIDS TO TREAT RESPIRATORY INFECTIONS AT THE AIRWAY INTERFACE ..... 38
*Annelise Barron*

(390D) THINKING SMALL FOR PANDEMIC SOLUTIONS: A NANOMEDICINE PLATFORM FOR DIAGNOSING AND TREATING VIRAL INFECTIONS ..... 39
*Thomas J. Webster*

**Author Index**

# 608379 Preventing Bacterial Colonization of Surfaces with Layer-By-Layer Coatings of Fluorinated Polyphosphazenes and Antibiotics

**Abstract:**

Preventing bacterial colonization of biomedical surfaces, ranging from skin grafts to hip implants, is highly desirable. Such a goal can be achieved by coating materials with biocompatible, polymeric layers that release antibiotics in response to a trigger that occurs at the onset of infection, such as locally induced acidification. Here, we explore pH-responsive coatings constructed with ionic fluorinated polyphosphazenes (PPzs). PPzs are promising building blocks for biomaterials due their known biocompatibility and unique structure of the repeat unit, which enables control over charge density and hydrophobicity as well as over the biodegradation rate. Traditional polyelectrolytes are unable to directly form coatings with cationic antibiotics at neutral pH and suffer from high degrees of antibiotic release upon exposure to physiological concentrations of salt. In contrast, fluorinated PPzs enabled unique, not possible with common polyelectrolytes, direct layer-by-layer assembly with cationic antibiotics, such as polymyxin B, colistin, gentamicin, and neomycin. The amount of antibiotics bound in the coatings could be tuned by the degree of fluorination of PPzs and/or film thickness. The coatings were able to prevent the onset of bacterial growth of *E. coli* and *S. aureus* in solution *via* antibiotic release (with antibiotic doses highly tunable *via* PPz fluorination degree and/or film thickness) and were also highly surface active. Importantly, coatings remained highly surface active against *E. coli* and *S. aureus* even after 30 days of pre-exposure to bacteria-free physiological conditions or after repeated bacterial challenge. Moreover, coatings displayed low (<1%) hemolytic activity for both rabbit and porcine blood. Deposition of the coatings on either hard (Si wafer, Ti substrate) or soft (electrospun fiber matrices) materials resulted in materials which were non-toxic towards fibroblasts (NIH/3T3) and displayed controllable fibroblast adhesion *via* PPz fluorination degree. Taken together, these results suggest a new way to form highly tunable, biocompatible polymer coatings for medical device surfaces.

# 607941 Treatment of Gram-Negative Bacteria Using Stimuli-Responsive Nanoantibiotics

**Abstract:**

In 2017, the World Health Organization confirmed that the world is running into a global crisis of antibiotic resistant bacteria. The clinical pipelines for antibiotics are few and even the chemicals in development are small modifications that are only short-term solutions to the problem We propose developing a platform using a unique combination of nanoparticles, conventional antibiotics and gene therapy to combat the rising antibiotic resistance problem. Our core technology is acid-transforming chitosan (ATC), a stimuli-responsive modification of chitosan that serves as the nanoantibiotic and delivery vehicle. We explored the antimicrobial capabilities of ATC and determined the possible mechanism for its antimicrobial efficacy. We determined the antimicrobial efficacy against various strains of three gram-negative pathogens (*E. coli*, *S. typhimurium*, and *P. aeruginosa*). These pathogens are commonly resistant to many antibiotics (beta-lactams, macrolides, tetracyclines, co-trimoxazole, and most fluoroquinolones), and has they have the capacity to develop resistance to a wide range of antimicrobials. ATC polymer was tested at varying pH conditions and we observed a positive correlation between pH and the polymer efficacy against the microbe. Additionally, ATC/fDNA polyplexes were applied to RAW 264.7 cells co-infected with S. typhimurium expressing GFP this resulted in a decrease in the levels of bacteria found within the cells as confirmed by flow cytometry. Furthermore, ATC/Ciprofloxacin nanoparticles were created by ionic gelation and their synergistic efficacy was observed. Overall, our results indicate that these nanoantibiotics provide an effective alternate strategy to treat bacterial infections.

# 605740 Nanosus (Ultra Fine Grained Stainless-Steel) for Orthopedic Implants

**Abstract:**

**Statement of Purpose:** Orthopedic implants are important therapeutic devices for the management of a wide range of orthopedic conditions. However, bacterial infections of orthopedic implants remain a major problem, and not an uncommon one, leading to an increased rate of osteomyelitis, sepsis, implant failure and dysfunction, etc. Treating these infections is more challenging as the causative organism protects itself by the production of a biofilm over the implant's surface (*1*). Infections start by the adhesion and colonization of pathogenic bacteria such as *Staphylococcus aureus* (SA), *Staphylococcus epidermidis* (SE), *Escherichia coli* (*E. coli*), Methicillin-Resistant *Staphylococcus aureus* (MRSA), and Multi-Drug Resistant *Escherichia coli* (MDR *E. coli*) on the implant's surfaces. Specifically, *Staphylococcus* comprises up to two-thirds of all pathogens involved in orthopedic implant infections (*2*). However, bacterial surface adhesion is a complex process influenced by several factors such as chemical composition, hydrophobicity, magnetization, surface charge, and surface roughness of the implant (*3*). Considering the intimate association between bacteria and the implant surface, we measured the effect of stainless-steel surface properties on bacterial surface attachment and subsequent formation of biofilms controlling above mentioned factors.

**Method:** The prominent bacteria responsible for orthopedic implant infections (SA, SE, *E. coli*, MRSA, and MDR *E. coli*) were used in this study. We were able to control the grain size of medical grade 304 and 316L stainless steel without altering their chemical composition (grain size range= 20 μm-200 nm) (*4*). Grain size control affected the nano-topography of the material surfaces which was measured by an Atomic Force Microscope (AFM). Grain sizes, such as 0.2, 0.5, 1, 2, 3, 9, and 10 μm, were used both polished and non-polished. All the stainless-steel samples were cleaned by treating with acetone and ethanol under sonication. Triplicates of all polished and non-polished samples with different grain sizes were subjected to magnetization of DM, 0.1 T, 0.5 T, and 1 T, before seeding them with the bacteria. Controls were used in the form of untreated samples. Bacterial were grown in Tryptic Soy Broth (TSB). An actively growing bacterial suspension was seeded onto the stainless-steel discs into 24-well micro-titer plates and kept for incubation. After 24 hours of incubation, the stainless-steel discs were washed with Phosphate Buffer Saline (PBS) to remove the plankton bacteria and allow the sessile bacteria in the biofilm to remain. The degree of development of the bacterial biofilms on the stainless-steel discs were measured using spectrophotometric analysis. For this, the bacterial biofilm was removed from the stainless steel by sonication. The formation of biofilms was also determined by performing a biofilm staining method using Safranin.

**Results:** AFM results revealed a slight decrease in roughness by decreasing the grain size of the material. Moreover, the samples were segregated into two categories of polished and non-polished samples, in which polishing decreased roughness significantly. After careful analysis we found out that polished surfaces showed a higher degree for biofilm formation in comparison to the non-polished ones. We also observed that bacteria showed a higher rate for biofilm formation for the demagnetized samples, whereas 0.5 T magnetization showed the least amount of biofilm formation. After 0.5 T, there was no significant change in the rate of biofilm formation on the stainless-steel samples. Altogether, stainless steel samples containing 0.5 μm and less grainsize, and magnetized with 0.5 T and stronger magnets demonstrated the least degree of biofilm formation.

**Conclusion:** In summary, the results demonstrate that controlling the grain size of medical grade stainless steel can control and mitigate bacterial responses on, and thus possibly infections of, orthopedic implants or other implantable devices. The research was funded by Komatsuseiki Kosakusho Co., Ltd (KSJ: Japan)

**References:**

1. M. Ribeiro, F. J. Monteiro, M. P. Ferraz, Infection of orthopedic implants with emphasis on bacterial adhesion process and techniques used in studying bacterial-material interactions. *Biomatter* **2**, 176-194 (2012).
2. M. Wang, T. Tang, Surface treatment strategies to combat implant-related infection from the beginning. *Journal of orthopaedic translation* **17**, 42-54 (2019).
3. I.-S. Yeo, H.-Y. Kim, K. S. Lim, J.-S. Han, Implant surface factors and bacterial adhesion: a review of the literature. *The International journal of artificial organs* **35**, 762-772 (2012).
4. S. Torizuka, E. Muramatsu, T. Komatsu, S. Nagayama, in *Nanostructured Metals and Alloys*. (Elsevier, 2011), pp. 715-746.

# 605802 B-Lactamase Responsive Hydrogels for Bacteria-Triggered Antibacterial Treatments

**Abstract:**

Antibiotic resistance is a major global threat increasing the difficulty of treating bacterial infections. Overuse and misuse of antibiotics has contributed to the rise in antibiotic resistance. There is a need for smart drug delivery systems that can self-regulate release of antibacterial agents to localize delivery, concentrate the drug at the site of infection, and reduce off-site toxicity to the microbiome, potentially lowering susceptibility to antibiotic resistance development. In this talk, I will describe our recent work on developing hydrogel technologies that respond to β-lactamases (βLs) (bacterial enzymes which are among the leading causes of antibiotic resistance) as a platform for bacteria-triggered release of encapsulated antibacterial agents.

We have functionalized a βL-cleavable compound with maleimides on both termini and utilized it as a crosslinker in the synthesis of responsive hydrogels through end-crosslinked polymerization with multi-arm thiol-terminated poly(ethylene glycol) (PEG) macromers via Michael-type addition. Bismaleimide PEG was used to form non-βL-degradable hydrogel controls. We observed that only hydrogels containing the βL-responsive moiety decreased in size and wet mass in the presence of βLs produced by three different bacterial species. The hydrogels degraded completely within 5 hours when incubated with 1 units(U)/mL βL from *Enterobacter cloacae*, 8 hours with 30 U/mL βL from *Pseudomonas aeruginosa*, and 9 hours with 400 U/mL with βL from *Bacillus cereus* (here 1 U is defined as hydrolyzing 1.0 μmol of benzylpenicillin per minute at pH 7.0 at 25°C). Control non-responsive hydrogels remained stable in βLs from all three species. Hydrogel degradation rates were further tuned by changing polymer density (complete degradation in ~13, 8, 3.5 hours for 15, 10, and 5% w/v hydrogels in 30 U/mL βL from *P. aeruginosa*, respectively).The release profile of fluorescent polystyrene nanoparticles (NPs) encapsulated within these hydrogels as model cargo tracked the βL-responsive hydrogel wet mass decrease rates, while no NPs were released from non-responsive hydrogels, indicating degradation-controlled cargo release. Hydrogels exhibited an on-off response when βLs were added or removed, respectively, further demonstrating their tunable βL-responsive behavior. The responsive hydrogels were found to degrade specifically when cultured with βL-producing strains of *P. aeruginosa* and *B. cereus* but remained stable in blank media or with a non-βL producing strain of *Staphylococcus aureus*. Again, the non-responsive hydrogels remained stable under all conditions and did not release any detectable NPs. To better mimic skin and soft tissue infections, we tested hydrogel degradation by βL-producing bacteria on infected agar or *ex vivo* porcine skin as opposed to shaking in broth and found similar results.

To demonstrate the versatility of this approach, we have also formulated supramolecular βL-responsive hydrogels with self-healing capabilities. These hydrogels were synthesized with the same βL-cleavable compound utilized in our PEG hydrogels, but this compound was now functionalized on both termini with adamantane moieties. The hydrogel was formed via non-covalent host-guest interactions of the adamantane moieties with polycarboxymethyl-β-cyclodextrins along with forming an interpenetrating network based on free-radical polymerization of acrylamide and *N*-vinylpyrrolidinone. As with the covalent PEG hydrogels, these hydrogels selectively degraded in the presence of βLs. They also have the added benefit of self-healing promoted by the host-guest interactions of the adamantane and cyclodextrin, which can increase their utility in environments subject to mechanical disruption. The βL-responsive hydrogel technologies that we developed have the potential to be used as prophylactic biomaterials, such as bandages or *in situ* forming injectable hydrogels, allowing for triggered therapeutic release, thereby improving management of infections, limiting exposure and reducing susceptibility to antibiotic resistance.

# 600968 Zwitterionic and Fluorinated Amphiphilic Polymer Coatings with Enhanced Resistance to Biofilm Formation

**Abstract:**

The pervasiveness of microbial fouling on biomedical interfaces that can cause serious infections has motivated research into antifouling materials that resist irreversible biofilm formation. Our efforts focus on amphiphilic polymer coatings produced via initiated chemical vapor deposition (iCVD) to present a surface energy landscape that reduces the adhesion of bacteria and biofilm formation. The present work details the iCVD copolymerization of hydrophobic 1*H*,1*H*,2*H*,2*H*-perfluorodecyl acrylate (PFDA) and 4-vinylpyridine (4VP) that is subsequently derivatized to a hydrophilic sulfobetaine zwitterionic moiety to form amphiphilic coatings. Synthesis of these zwitterionic-fluorinated amphiphilic copolymers is uniquely enabled by the all-dry nature of the iCVD technique. Individually, PPFDA coatings exhibit significant promotion of *Pseudomonas aeruginosa* biofilm growth, while zwitterionic polymer coatings reduce biofilm formation in accordance with existing studies of antifouling zwitterionic materials. In conjunction, the random copolymerization of these two species with drastically differing surface energies displays enhanced reduction of biofilm formation in the short term. In this work, we report the synthesis and characterization of this new amphiphilic polymer coating and assess its superior ability to reduce the adhesion of biofilm-forming *P. aeruginosa*.

# 603608 Inhibiting Bacterial Surface Colonization Using Peptoid-Loaded Polyelectrolyte Microgels

**Abstract:**

Antimicrobial modifications of biomaterial surfaces have been explored extensively as a means to prevent infection associated with tissue-contacting biomedical devices. As an alternative to traditional antibiotic elution-based delivery methods, which can promote the development of antimicrobial resistance, we have been exploring an approach where polyanionic microgels are deposited on biomaterials surfaces and cationic antimicrobials are sequestered within them for many weeks by complexation. Nevertheless, when challenged by bacteria the antimicrobial can be released and provide for local microbial killing, a process that appears to be driven by the high negative charge and hydrophobicity associated with the bacterial cell envelope [1]. For this so-called self-defensive behavior [2] to be successful, the complexation strength must be such that the antimicrobial can remain sequestered for extended periods of time in physiological medium that contains various salt species, all of which are able to interfere with the electrostatic complexation. We have previously shown that poly(acrylic acid) (PAA) can sequester colistin, an FDA-approved antibiotic with +5 net charge, in low ionic strength buffer for weeks. Increasing the ionic strength, however, screens the colistin-PAA complexation interaction and enables rapid burst release of the colistin. In contrast, we have identified particular antimicrobial peptides that remain stably sequestered in PAA microgels exposed to buffers with physiological ionic strength for extended periods of time. To better understand the key antimicrobial characteristics that influence complexation, we have been using model peptoids whose size, charge, hydrophobicity, and aromaticity can all be systematically varied. Peptoids are *N*-substituted glycines whose side chains are appended to the nitrogen atom rather than to the α-carbon of an amino acid. Much of our initial work has concentrated on peptoid TM1 ($M_w$ = 1819.36 Da, +5 net charge), a helical 12mer which is an effective antimicrobial. We complex it with PAA microgels synthesized by membrane emulsification and UV photopolymerization. We use *in situ* optical microscopy to measure changes in the diameter of microgels immobilized on glass substrates as a means to follow peptoid loading, sequestration, and release behavior. We find that the average microgel diameter decreases by a factor of about 40% when exposed to TM1-containing low-ionic strength phosphate buffer. The diameter increases minimally over one week when the ionic strength is increased to 0.14 M, indicating that the peptoid remains sequestered. Using a PetriFilm assay we furthermore show that TM1-loaded microgel-modified glass substrates resist colonization by staphylococcal bacteria. Varying from the basic TM1 motif, we have studied a series of other peptoids, including TM22 which shares the same net charge but replaces aromatic *N*-S-phenylethyl moieties found in TM1 with *N*-S-*sec*-butyl groups. We find stable complexation even when the net charge is lowered provided that aromaticity remains, clearly indicating that π-cation, π-anion, and/or π-π interactions are influencing the complexation strength in a manner less sensitive to salt screening than the electrostatic interactions. More generally, these findings can help develop design principles to guide the choice of microgel chemistry and possible complexed antimicrobials available to create self-defensive surfaces able to simultaneously promote healing and inhibit infection.

[1] J. Liang, H. Wang, and M. Libera, *Biomaterial surfaces self-defensive against bacteria by contact transfer of antimicrobials.* Biomaterials, 2019. **204**(June): p. 25-35.

[2] X. Xiao, W. Zhao, J. Liang, K. Sauer, and M. Libera, *Self-Defensive Antimicrobial Biomaterial Surfaces.* Colloids and Surfaces B: Biointerfaces, 2020. **Available online 21 April, 2020 (**

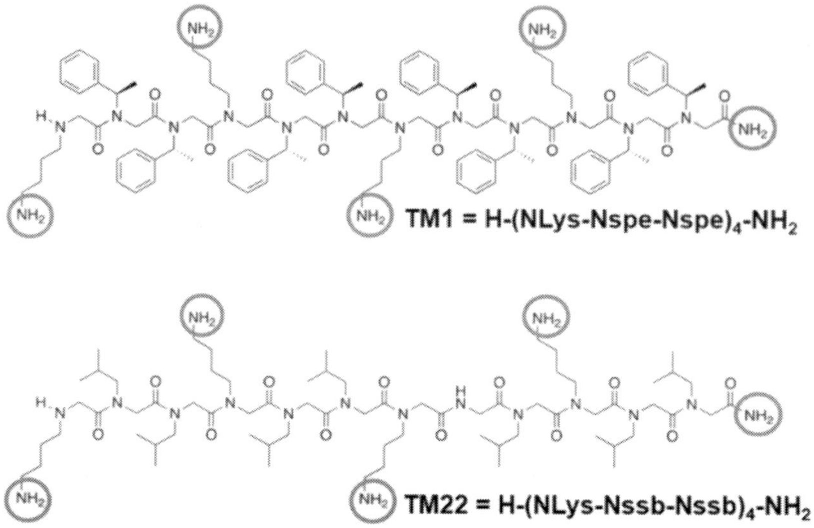


Figure 1 – Chemical Structures of Peptoid TM1 (Top) and TM22 (bottom) with the same electrostatic charge (+5), differ by the presence/absence of aromatic moieties in the uncharged residues.

# 604435 Fabrication of Oil-Infused Anti-Biofouling Coatings on the Surfaces of Flexible Polymer Tubing

**Abstract:**

Advances toward the design of liquid-infused surfaces (LIS) or slippery liquid-infused porous surfaces (SLIPS)—materials fabricated by the infusion of oily lubricants into chemically compatible porous or textured surfaces—have led to new classes of synthetic materials with robust and mechanically-compliant anti-fouling properties. Several reports demonstrate that these materials can resist adhesion and colonization by microorganisms and prevent formation of fungal or bacterial biofilms in both static and flowing media. Here, we report strategies for the fabrication of anti-biofouling SLIPS on the outside and inside (luminal) surfaces of flexible polymer tubing typically used to design catheters or transport fluids in commercial and industrial contexts. Our approach is based on the infusion of hydrophobic oils into rough and nanoporous polymer matrices fabricated by reactive/covalent layer-by-layer assembly. These methods can be used to fabricate slippery coatings on the surfaces of flexible tubing up to one meter in length. These coatings retain their antifouling properties upon repeated flexing, bending, and coiling, and after (i) ethylene oxide sterilization, (ii) contact with blood or simulated urine, and (iii) exposure to flow conditions for periods of weeks to months. Our results reveal these SLIPS-coated tubes to prevent the formation of bacterial biofilms (>95% relative to uncoated controls) upon exposure to cultures of *Staphylococcus aureus* for up to seven days, and that the anti-biofouling properties of these tubes can be improved and prolonged by adopting additional strategies that permit the sustained release of broad-spectrum antimicrobial agents. These approaches are scalable and could provide new anti-biofouling solutions for a broad range of applications, including clinical (catheters) and industrial (food processing) scenarios in which the fouling of flexible tubing is endemic. Results highlighting the potential of these materials in these applied contexts will be discussed.

# 610995 Modifying Plastic Surfaces in Dairy Membranes to Act As Antimicrobial Surfaces

**Abstract:**

Downstream contamination of dairy processes by Listeria monocytogenes (LM) is a serious issue because of the associated mortality rate in immune-compromised individuals. Dairy membranes can be modified to include solid-phase antimicrobials to kill or mitigate growth of LM. The feed spacer mesh of spiral wound dairy membranes has been functionalized with Quaternary Amine Compounds (QACs), a family of broad spectrum antimicrobials that are effective against Gram-positive bacteria. The antimicrobial is covalently bound to the polypropylene surface to avoid leaching into the product stream. Growth curve studies with a surrogate have demonstrated the effectiveness of the surface, and well as the necessary robustness to withstand clean-in-place conditions that will be required when implemented in industry. We have also functionalized the surfaces with polymerized phenolic compounds and comparisons were made with QACs. Preliminary results will also be presented in milk and brine systems.

# 597214 Interfacial Protein Self-Assembly: An "Ironic" Approach Towards Generating Antifouling Coatings

**Abstract:**

Non-specific adsorption of proteins, also known as fouling, is a "common but very complicated phenomenon" hallmarked by its ubiquity. Most research aims to prevent fouling, which can be detrimental to the function of devices, such as biosensors and membranes, and can promote harmful pathologies, such as antibiotic-resistant infections. Rather than regarding fouling as a persistent problem, our work leverages the non-specific interactions of silk fibroin during self-assembly to form stable and substrate-tolerant thin-film coatings. These coatings provide complete surface coverage and can grow to tens of nanometers thick, transforming the physicochemical properties of a surface in a one-pot aqueous process without requiring substrate pre-activation or covalent chemistries. Our studies have demonstrated that silk fibroin coatings can be formed on a variety of substrates ranging from hydrophobic Teflon to hydrophilic $TiO_2$. Furthermore, these coatings can exhibit beneficial properties for resisting the adhesion of gram negative and gram positive bacteria. Our research moreover delves into the complex interplay of surface-protein and protein-protein interactions underlying coating formation to establish methods by which we can tune the coating process.

# 603794 Correlating *Staphylococcus epidermidis* biofilm pH Gradients with Local Biofilm Mechanical Properties to Reveal Contributions of pH to Biofilm Disassembly and Dispersion

**Abstract:**

Bacterial biofilms are surface-adherent communities of microbial cells encased within a protective extracellular polymeric matrix that are frequently responsible for device-associated and chronic infections. The biofilm lifecycle consists of the following stages: initial bacterial surface adhesion, accumulation of bacteria and matrix materials on a surface, maturation into an organized community, and disassembly and dispersion of cells from the surface. The final stage of the biofilm lifecycle—disassembly and dispersion—contributes to the spread of infection within the body. The heterogeneity of the structural and mechanical properties of biofilms can result in the formation of pH gradients within biofilms due to the nonuniform distribution of nutrients or metabolic waste products. Bulk changes to the pH of staphylococcal biofilms above 7.3 have been shown to soften staphylococcal biofilm matrix materials[1]. However, the extent to which biofilms can naturally control their mechanics through modulating pH is not well understood. Here we correlate microscale pH maps with biofilm mechanics to determine the role of natural pH gradients in controlling *Staphylococcus epidermidis* biofilm mechanics and disassembly. We use confocal laser scanning microscopy (CLSM) and image analysis techniques to quantitatively map pH values and determine the local mechanical behavior of biofilms using passive microrheology. We find that *S. epidermidis* biofilms grown for 24 hours contain more acidic pH regimes than those grown for 6, 12, and 18 hours. Spatially, we observe the most acidic microenvironments to be located close to the biofilm-substrate interface. Additionally, local *S. epidermidis* biofilm mechanics are found to vary spatially with pH, where regions of the biofilm with higher pH ($pH > 6.9$) are mechanically softer than regions of the biofilm with lower pH ($pH < 6.9$). This work advances the understanding of how natural biofilm pH gradients contribute to biofilm mechanics and reveals potential contributions of matrix phase stability to the disassembly and dispersion of staphylococcal biofilms.

[1]Stewart, E.J. et al. *Scientific Reports,* (2015).

# 611762 Thermally Triggered Biofilm Dispersion

**Abstract:**

Biofilm infections on medical implants are a multi-billion dollar problem afflicting hundreds of thousands of patients in the U.S. each year. They cannot be eliminated by antibiotics, so strategies to replace the current standard of care—explantation and replacement—have attracted significant interest. By applying a localized thermal shock, we have shown that these biofilms can be reduced and eliminated, though the mechanism of elimination has been unclear. Moreover, we found that applying antibiotics during the thermal shock, dramatically reduced the required time and temperature to eliminate biofilms infections of both *Pseudomonas aeruginosa* and *Staphylococcus aureus* cultured using two different protocols (shaker table, drip flow reactor). One hypothesis for this synergistic efficacy is that the thermal shock triggers a change in bacterial phenotype, prompting the bacteria to disperse from the biofilm in the planktonic state, and thereby making them susceptible to antibiotic elimination.

Flow cell (1 x 1 x 6 inches) with mature *Pseudomonas aeruginosa* biofilms placed at the bottom and a constant flow of diluted TSB at room temperature was used to determine the bacterial dispersion rate from the biofilms. We did high flow rate trials at 800 ml $min^{-1}$, and we learned that bacterial dispersion is so high that despite the large volume of effluent, dispersed bacteria concentrations are still large enough for quantification, but there's also concern about debridement even without elevated temperature. Further trials at a lower flow rate 80 ml $min^{-1}$ established a baseline dispersion event with little observed debridement, and subsequent in situ thermal shock at 70°C for 5 min triggered a spike in effluent bacteria. This supports our hypothesis thermal shock promote biofilm bacterial dispersion to planktonic state which is more susceptible to antibiotics and suggests that biofilms on medical implant devices can be mitigated by combining localized thermal shock with antibiotics.

# 600999 Biointerfacial Design to Manipulate Siderophore Production By *Pseudomonas Aeruginosa* without Soluble Chemical Intervention

**Abstract:**

Ample biointerfacial materials research has focused on biocidal or antifouling outcomes, yet some microbes can be utilized for advantageous qualities when allowed to colonize a surface under appropriate conditions. Our approach to harnessing these benefits involves the use of initiated chemical vapor deposition (iCVD) to design polymer thin film coatings with properties that control the behavior of exposed bacteria. To uptake iron, an important life-sustaining element, a variety of microbes produce a secondary metabolite called a siderophore that has a high binding affinity for iron. The dynamics of this diffusive scavenging technique has a profound impact on competition between minute species in a shared ecosystem. The present work examines iCVD-generated polymer thin films that behave cooperatively in iron sequestration and change the expression of siderophores and biofilm formation by bacteria that contact surfaces coated with these films. The material characterized herein sets the stage for future work on tunable biointerfaces that influence microbial activity exclusively through surface phenomena without reliance on soluble elements.

# 604393 Controlling the Adhesion of *Staphylococcus Aureus* to Polydimethylsiloxane Biomaterials

**Abstract:**

Bacterial adhesion and consequent biofilm formation are recurrent problems that lead to infections and death. Antibiotics, the standard but imperfect means of combatting these infections, kill bacteria while also contributing to antibiotic-resistant bacteria. In particular, the growing prevalence of infections caused by methicillin-resistant *Staphylococcus aureus* (MSRA) has directed our work to better understand the initial adhesion of *S. aureus* to medically relevant biomaterials and potentially, develop alternative approaches that can be employed to reduce bacterial adhesion. The mechanical properties and chemistry of hydrogels, such as stiffness and hydrophilicity, respectively, have been demonstrated to affect the initial rate of bacterial adhesion to different surfaces. However, more questions remain as to the attachment of microbes to polydimethylsiloxane (PDMS), a common polymeric organosilicon compound used for medical devices and coatings. PDMS was fabricated into gels that had various mechanical properties and concentrations of unreacted oligomers. Material properties of the gels, such as tackiness, viscoelasticity, homogeneity, and hydration, were quantified using rheology, Fourier-transform infrared spectroscopy, and contact angle measurements. To better understand the mechanism of attachment, the adherence of *S. aureus* ATCC 12600, a genome-sequenced strain, and *S. aureus* SH1000, a clinically-isolated strain, to the PDMS gels was assessed using a 24 hour assay. This work provides insight into the mechanism of adhesion for *S. aureus* to PDMS gels and potentially, also how to design bacteria-resistant surfaces for a broad range of applications, including medical implants.

# 611614 Mass-Transfer Relations for for Nanoparticle Transport across Human-Biofilm Interface

**Abstract:**

A set of current issues surrounding implanted medical devices are the bacterial infections which occur due to formation of biofilms. Nanoparticles have been proposed as therapeutics for biofilms due to their capacity to penetrate biofilm ECMs and deliver drugs to the cell surface.[1] Pharmacokinetic models for these therapeutics will need mass-transfer relations for these human-biofilm interfaces to quantify the accumulation of nanoparticles in the biofilm. A first-principles model for nanoparticle transport in biofilms was developed which accounts for nanoparticle diffusion and macro/microscale electrostatic interactions within biofilms. Nanoparticle diffusion and macroscale electrostatic interactions were captured using a modified Nernst-Plank equation. Microscale electrostatic interactions were accounted for using Negative Hill Coefficient Binding Kinetics. Dimensionless mass-transfer relations were developed from this model for different rate-limiting assumptions by solving the general component balance using both analytical and numerical techniques. These relations were then applied to charged quantum dots diffusing in *B. subtilis* biofilms to predict qualitative behaviors of this system.

# 607550 Salivary Mucin Glycans Suppress Quorum Sensing and Associated Virulence Traits in *Streptococcus Mutans*

**Abstract:**

Mucus barriers accommodate hundreds of trillions of microbes throughout the human body, while preventing pathogenic colonization. In the oral cavity for instance, saliva containing the salivary mucins MUC5B and MUC7 forms a slippery mucosal pellicle that coats the soft tissue and teeth to prevent infection by oral pathogens. Previous studies have found salivary mucins can interact directly with microbes through selective agglutinin activity and bacterial binding, but the extent of saliva's protective functions and their underlying mechanisms are not well understood. Here, we identify that native saliva, and specifically, its gel-forming mucin polymer MUC5B, dramatically inhibits infection-related traits in a clinically-important oral pathogen, Streptococcus mutans. In particular, we find MUC5B overrides bacterial communication by suppressing key quorum sensing pathways regulated by the Competence Stimulating Peptide (CSP) and the SigX Inducing Peptide (XIP). These quorum sensing signaling systems regulate the expression of virulence factors that enable infection of the host. We show that MUC5B suppresses two of these problematic traits, namely, bacteriocin production and the acquisition of competence, a prerequisite for natural genetic transformation. Our data reveal that MUC5B's suppression of quorum sensing is mediated by mucin glycans. We find that these glycans are highly effective at suppressing quorum sensing and the acquisition of competence even when removed from the mucin backbone. Together, these results suggest that mucin-type glycans represent highly attractive compounds for the prevention of S. mutans virulence in the oral cavity.

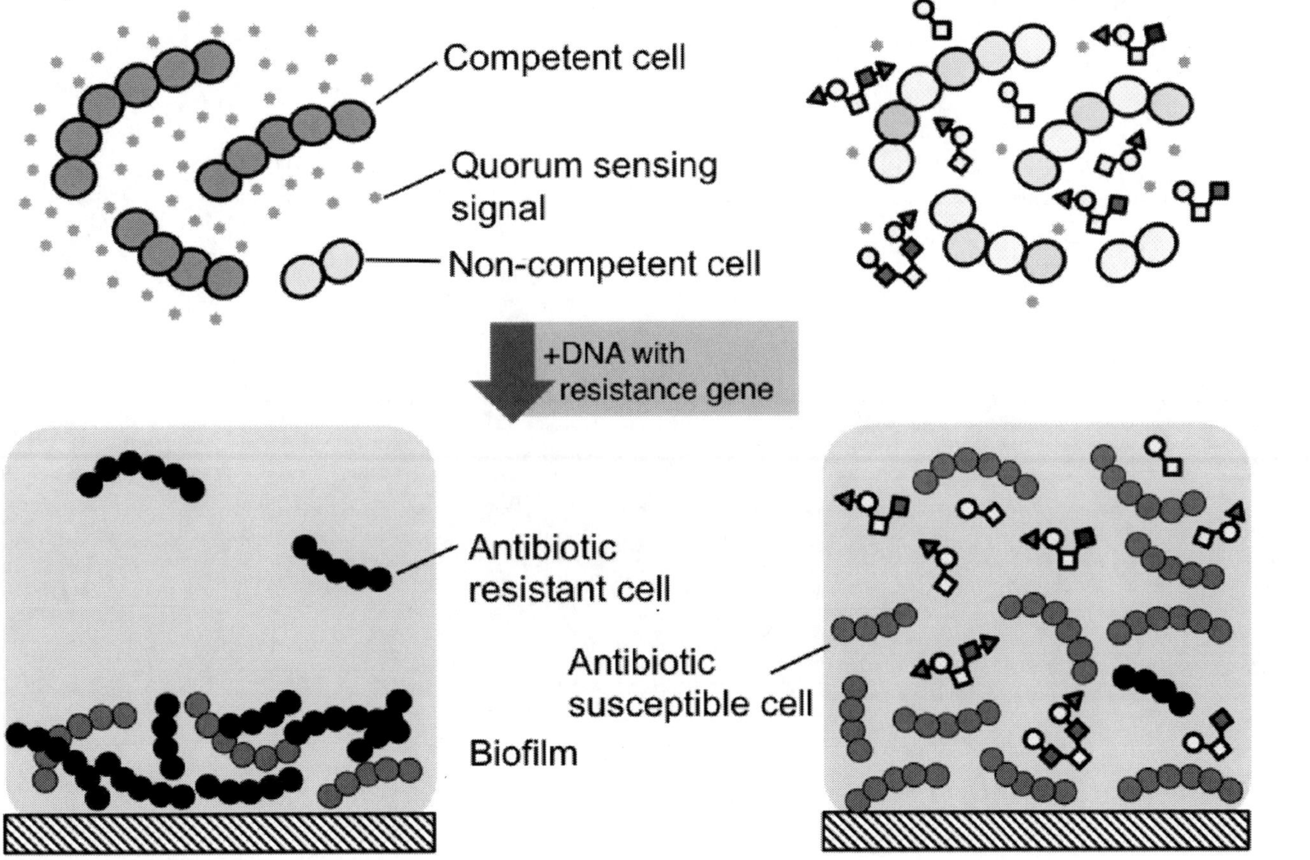

# 606270 Characterization of Interfacial Films of Mucoid and Nonmucoid *Pseudomonas Aeruginosa* Isolates.

**Abstract:**

Fluid interfaces are energy rich environments ideal for the self-assembly of micro/nano sized particles and film formation by biological entities. The formation of such films appears to be a dynamics process, in which cells respond to the entrapment and the existing interfacial energies by forming a film appropriate to their survival. We hypothesize that cells subjected to continuous interfacial stress adapt and evolve to exhibit phenotypic changes essential to their survival. In this study, we investigated the interfacial properties of films formed by non-mucoid (PANT) and mucoid (PASL) strains of *P. aeruginosa* isolated from CF (Cystic Fibrosis) patients. Under constant strain, PANT forms a conserved viscoelastic film exhibiting similar response under compression and tension whereas PASL exhibits more relaxation under compression than tension. Microscopic analysis revealed that interfacial films of PASL retained an intact cellular morphology compared to PANT, which include large number of lysed cells. This indicates that mucoid coating of PASL plays an important role in protecting the bacteria from interfacial stresses. Transcriptional profiling demonstrated that the gene expression under interfacial confinement is distinct to mucoid and non-mucoid cells suggesting that phenotypic differences in cells alter their behavior at fluid interfaces.

# 604127 Nanotopographical Amplification of Bacteria-Material Repulsive Interactions for Mitigating Bacterial Attachment and Biofilm Formation

**Abstract:**

Pathogenic biofilms persisting on abiotic surfaces have wreaked havoc in healthcare and food safety. Numerous practical experiences and researches have made evident that the eradication of mature bacterial biofilms is extremely difficult with the conventional removal and disinfection methods, including the use of antibiotics. Recently, preventative strategies targeting the most vulnerable step of the biofilm life cycle – i.e. the reversible attachment of planktonic cells to abiotic surfaces – have shown particular promise. Here we present a novel nanotopography-enabled strategy, which leverages the repulsive cell-substrate interaction forces, to inhibit bacterial attachment and biofilm formation.

To this end, anodized aluminum oxide (AAO) substrates with defined pore diameters (10nm, 25nm, 50nm, and 100nm) and a smooth control, were incubated statically for 48h with pure cultures of *Escherichia coli*, *Listeria monocytogenes*, *Staphylococcus aureus*, and *Staphylococcus epidermidis*. Surface properties (surface charge and surface energy) and morphology (or topography) of the test strains and the AAO substrates were characterized using contact angles, zeta potential measurements, SEM, and AFM. These results were used as inputs for predicting the cell–substrate interaction forces, based on the extended Derjaguin–Landau–Verwey–Overbeek (XDLVO) theory. The force predictions were correlated with the surface-bound biomass (quantified by confocal laser scanning microscopy) for all strain-topography pairs, to evaluate the predictive power of the XDLVO model. Furthermore, force-distance curves were collected using functionalized AFM colloidal probes to directly verify the nanotopography dependency of the repulsive forces.

AAOs with pore diameters of 15nm and 25nm were able to effectively reduce bacterial attachment and biofilm formation by all strains compared to the smooth control; however, 50nm- and 100nm-pore AAOs resulted in more biomass accumulation. Logarithmic biomass was found negative correlated with the maximum repulsive forces ($F_{max}$) predicated by the XDLVO model ($r = 0.79$). The main contributors to $F_{max}$ were the surface-originated electrostatic and acid-base interactions, which were substantially enhanced by the immense vertical surface area characteristic of the 15nm- and 25nm-pore AAOs. Furthermore, AFM force-distance measurements suggested that cellular appendages and the membrane bound polymeric molecules also played an important role in the nanotopography dependency of the reversible attachment.

These findings elucidated the nanotopography dependency of bacterial attachment and biofilm formation, which in turn serves as a roadmap toward more effective bacteria-repellent nanotopographies for mitigating pathogenic biofilms.

# 602404 Interaction of Cyanobacteria with Nanometer and Micron Sized Polystyrene Particles in Marine and Fresh Water


**Abstract:**

Microplastics and nanoplastics are emerging pollutants, widespread both in marine and freshwater environments. Cyanobacteria are also ubiquitous in water and play a vital role in natural ecosystems, using photosynthesis to produce oxygen. Using photography, fluorescence microscopy and cryogenic and scanning electron microscopy (cryo-SEM, SEM) we investigated the physicochemical response of one of the most predominant seawater cyanobacteria (*Synechococcus elongatus PCC 7002),* and freshwater cyanobacteria (*Synechococcus elongatus Nageli PCC 7942*) when exposed to 10 mm diameter polystyrene (microPS) and 100 nm diameter polystyrene (nanoPS) particles. Marine and freshwater cyanobacteria formed aggregates with the nanoPS, bound together by extracellular polymeric substances (EPS), and these aggregates sedimented. The aggregates were larger, and the sedimentation was more rapid for the marine system. Aggregate morphologies were qualitatively different for the microPS samples, with the bacteria linking up a small number of particles, all held together by EPS. There was no sedimentation in these samples. The cyanobacteria remained alive after exposure to the particles. The particle size- and salt concentration-dependent response of cyanobacteria to these anthropogenic stressors is an important factor to consider for a proper understanding of the fate of the particles as well as the bacteria.

# 611620 Using Freely Suspended Biofilms to Study Interactions Among Bacterial Cells: Mechanical and Transport Properties

**Abstract:**

Biofilms - communities of bacterial cells encased in their extracellular polymeric matrix - are relevant materials in both medicine and engineering: the same features of a biofilm that often pose challenges to its eradication in a medical setting (three dimensional structure, cohesiveness, and resilience in the face of environmental changes) can be beneficial to a deliberately designed structural material.

In either case, there is a need to measure the physical properties of biofilms at a practically relevant (bulk) scale. However, measuring bulk scale features is complicated by the fact that biofilms are inherently heterogeneous materials, and their mechanical behavior is also closely linked to microscopic features: the chemistry of cell-cell interactions, local composition, and structural heterogeneity. Existing methods to probe biofilms at a bulk scale do not always take these different length scales into account, sometimes using a local section of biofilm as a proxy for the behavior of the whole film, or neglecting structural heterogeneity by pooling many plates' worth of material for a single measurement.

We try to address these limitations by making custom millifluidic devices that suspend a three millimeter biofilm sample across a support, and impose tunable hydrostatic pressure drops in the Pa-kPa range across the film. The resulting deformation of the film through an aperture is visualized with optical coherence tomography and used to estimate bulk mechanical properties of the film. Our method requires only microliters of material, causes minimal disruption to the film structure, allows for measurement of both average properties as well as local heterogeneity, and is compatible with other types of measurements as well (e.g. confocal microscopy, perfusion, or optical properties).

We pair our device with a model system: *E. coli* K-12 DH10B, an inherently poor biofilm former, engineered to surface display membrane-tethered elastin like polypeptides (ELPs) with various repeat lengths. These precisely tailored building blocks then grow and assemble into bulk materials, which we test. We investigate the effects of varying ELP repeats and presence or absence of a disulfide bond-forming cysteine on the bulk mechanical properties of the biofilms.

# 604397 Zwitterion-Containing Copolymers Reduce Bacterial Adhesion in Dental Environments

**Abstract:**

Bacterial attachment is a consistent problem affecting a wide range of biomedical interfaces, including wound healing and dental infections. The decreased use of antibacterial agents necessitates more biocompatible, antifouling polymer coatings that do not spread antibiotic-resistance genes. Here, we report the synthesis of new copolymers that formed antifouling coatings on multiple surfaces, including porous hydroxyapatite. Specifically, we systematically investigated how altering the composition of copolymers containing catechol methacrylate-poly(methacryloxyloxyethyl phosphorylcholine) and catechol methacrylate-polyMPC-methacrylic acid impacted their adhesive and antifouling properties, allowing for the smart design of functional coatings. Polymer coatings were characterized using ellipsometry, contact angle, and x-ray photoelectron spectroscopy, revealing that these copolymers formed uniform, hydrophilic coatings that were 10 nanometers-thin. Notably, the copolymer coatings self-adhered to hydroxyapatite and reduced the attachment of *Escherichia coli* and *Streptococcus oralis* without killing the microorganisms. Additionally, experiments that mimicked the mouth environment (i.e., swallowing) were employed to evaluate *Streptococcus oralis* adhesion, which was reduced and deemed reversible adhesion by these copolymer coatings. This study evaluated the effect of varying the amount of adhesive and antifouling groups on the efficacy of in-house synthesized antifouling coatings. We suggest that these novel copolymers could enable the smart design of antifouling surface coatings for medical and dental applications.

# 601000 Novel Coating Material for Controlling Corrosion and Biofouling in Water Supply System

**Abstract:**

Corrosion and microbial contamination in water supply system could cause fouling and failure of drinking water pipes. Microorganisms colonize on pipe surfaces by forming biofilm that resists chlorine disinfectant, which could lead to reduction of distribution efficiency, shorter service time but higher energy consumption. Physical and biological fouling in the drinking water supply system not only remarkably increase the water treatment cost, but also seriously threaten water quality and human health. This research project is to develop safe and effective anti-corrosion and anti-biofouling coating material for water pipes and pumps. The new coating is composed of epoxy matrix and antimicrobial polymers including polyethyleneimine (PEI) and polyhexamethylene biguanide (PHMB). Studies show that the new coating has excellent adhesion property on metal substrates, as well as exhibits contact-killing and bactericidal activities to effectively prevent biofilm formation on different substrates including metal, glass and plastics. Moreover, a pilot-scale friction test rig was installed to simulate the real use condition of the coating material. Compare to uncoated water pipes and pump, the new coating material endows pipes with a highly smooth surface. It could obviously decrease friction loss of pipes while enhancing pump efficiency. More importantly, it can avoid the formation of biofilm even after long-term test. Besides, ageing acceleration test indicates the coating could be stably operated for years. The application of new coating material would focus on safeguarding water distribution system, while simultaneously improving their efficiency with decreasing in cost, energy consumption and greenhouse gas emission associated with water supply.

# 600090 Direct Imaging Methods Show Limitations of Traditional Biofilm Staining Techniques

**Abstract:**

Biofilm research is of great importance to better understand the microorganisms that pose major threats to food safety, security and public health. Biofilms are most frequently characterized using staining techniques in a well plate. The inherent drawbacks of traditional staining assays invite the development of orthogonal techniques to assess biofilm formation. In this work, we present a set of simple and direct imaging methods to detect biofilms at both solid-liquid and air-liquid interfaces in four bacterial species with a diverse range of biofilm forming abilities. By cultivating biofilms at an air-liquid and particle laden air-liquid interface in a well plate and a pendant droplet, these non-traditional imaging methods highlight false positives for biofilm detection using the traditional crystal violet staining assay.

# 596877 Multifunctional Nanoculture System for Studying Microbial Dynamics

**Abstract:**

The diversity and abundance of species in the human microbiome have significant implications in the disease process. For instance, microbial communities in healthy mouth and gastrointestinal (GI) habitats can be affected by microenvironmental changes (e.g. temperature, humidity, pH, and chemical composition) that lead to unbalanced interactions promoting the growth of pathogens with consequences including periodontal diseases, C difficile infection, cancer, sepsis, and death. However, tools to characterize and model the signatures of healthy or diseased human microbiomes are lacking, in part due to the inability to culture some species implicated in such dynamics or define the exact cell interactions. Here, we propose to develop synthetic communities modeling human microbiome. Using a high-throughput microfluidic-based technique, we generate microbial communities in artificial microniches (nanocultures) to grow microbes with challenging culturability. Each nanoculture begins as a several nanoliter droplets of suspended cells, encapsulated by a polydimethylsiloxane (PDMS) membranes. The PDMS shell provides long-lasting mechanical support, and transport properties enabling long term study of microbial dynamics in controlled microenvironments. The resulting system is used to investigate the physicochemical interactions involved in microbial pathophysiology, microbial persistence, and antibiotic resistance.

# 601530 Understanding the Role of Protein Deposition in Microbial Interactions with Catheters Using Liquid-Infused Surfaces

**Abstract:**

The treatment and control of catheter-associated urinary tract infection (CAUTI) is becoming increasingly challenging due to the rise of antibiotic-resistant pathogens. Part of the problem lies in the fact that the colonization of catheter surfaces is a complex interaction between the pathogen, the surface, and the host that is difficult to untangle. For example, recent work has shown that fibrinogen (Fg) is released during catheterization and deposited on the catheter surface, coating it and forming a platform for colonization by CAUTI-associated pathogens. Yet the extent of the role it plays remains unclear. In this work, we use bio-inspired liquid-infused surfaces fabricated on catheter materials to assess the contribution of proteins in the deposition and formation of biofilms both *in vitro* and *in vivo*. We find that the amount of protein on the surface of the catheter material can be adjusted through modulating the amount of liquid presented at the surface. This in turn affects the interaction of the bacteria with the surface, with more bacteria colonizing surfaces where more protein is present. The results demonstrate that liquid-infused surfaces can be a useful tool in untangling complex interactions in the colonization of catheter surfaces by bacterial pathogens in CAUTI.

# 605908 Revealing Genetic Functions during Bacterial Adhesion and Biofilm Formation Using Barcode Transposon Mutant Library

**Abstract:**

Many bacteria live associated with material surfaces. This sessile mode of bacterial cells is phenotypically different from their planktonic counterparts, and often lead to chronic infections in humans and biofouling in water pipes. However, the genetic functions and regulations in bacterial adhesion and biofilm formation remain largely unknown. There are multiple difficulties with gaining a comprehensive molecular understanding of these processes. First, many genes are involved in bacterial adhesion and biofilm formation, and the regulatory pathways change at different steps. Second, the genetic functions are dependent on the genetic background of the host strain, the material surface properties, and environmental factors. Thus, the bacterial adhesion and biofilm formation of one bacteria on one material are often not the same as the bacterial adhesion and biofilm formation of another bacteria on another material. To have a systematic understanding of bacterial adhesion and biofilm formation, a high-throughput assay is required. Here, we developed a high-throughput screening assay to reveal the functions of 4000 genes during bacterial adhesion and biofilm formation in 2 days, using a barcoded transposon mutant library.

To demonstrate the method, we used an *E. coli* K-12 barcoded transposon mutant library as the model strain and polydimethylsiloxane (PDMS) as model material to test bacterial adhesion and biofilm formation under various conditions. By tracking the mutation fitness on surfaces, this method successfully identified the overall genetic functions in different steps of biofilm formation. For example, lipopolysaccharides (LPS) synthesis pathway and TolA-TolQ-TolR complex are important for the bacterial adhesion. In particular, short LPS decreases the bacterial adhesion. These results are consistent with previous reports based on both the classical Derjaguin-Landau-Verwey-Overbeek theory prediction and the experimental observation. Moreover, the genetic functions of early stage biofilm formation are different from those during adhesion. Type I fimbriae and flagella are the key players for *E. coli* staying and growing on PDMS during early biofilm formation. Loss of type I fimbriae causes low biofilm formation compared to wild-type cells; fimbriae promotion significantly increases the biofilm formation. But this phenomenon does not occur in the bacterial adhesion. Similarly, although the mutations of LPS and TolA-TolQ-TolR complex exhibit defects in adhesion, most of them do not show defects in early biofilm formation, highlighting that genetic functions change during different processes and suggesting that bacterial adhesion inhibition may not be successful to inhibit biofilm formation. In addition to these genes, multiple hypothetical proteins were identified to influence biofilm formation as well.

This high-throughput method is a powerful tool for systematically revealing the genetic functions in bacterial adhesion and early biofilm formation. Ultimately, by measuring the genetic functions in various bacteria, materials, and environmental factors, we expect to discover the fundamental principles of bacterial adhesion and biofilm formation.

# 604084 Elucidating a New Extracellular Function of Degp Inhibiting Biofilm Formation of Enterohemorrhagic *Escherichia coli* O157:H7

**Abstract:**

Enterohemorrhagic *Escherichia coli* O157:H7 (EHEC) is a pathogen causing serious outbreaks of hemorrhagic colitis. Conventional antibiotic treatment is not recommended for EHEC infection as antibiotics trigger Shiga-toxin production of EHEC and aggravate hemolytic-uremic syndrome. Biofilm formation of EHEC is closely associated with its virulence expression. Previously, we identified that probiotic *Escherichia coli* Nissle 1917 (EcN) secretes DegP and thereby inhibits EHEC biofilm formation. However, the extracellular function of DegP inhibiting EHEC biofilms is unclear. DegP is known as a serine protease exhibiting both proteolytic and chaperone functions and binds to outer membrane proteins (OMPs) of target cells. We hypothesized that DegP interacting with OMPs of EHEC may inhibit EHEC biofilm formation. We constructed EHEC mutants lacking *ompA*, *ompC*, or *ompF* individually or in combination and assessed their biofilm formation in the presence of DegP-secreting EcN in the co-culture or by adding purified DegP. When the *omp* were deleted in EHEC strains, their biofilm inhibition by DegP-secreting EcN was lessened, indicating that DegP may interact with OMPs to repress EHEC biofilm. Similar results were obtained when purified DegP was added into EHEC culture to form single biofilms. Next, we examined whether the EHEC biofilm inhibition is caused by protease activity or chaperone activity of DegP. A DegP variant (DegP S210A) with protease function disabled still inhibited EHEC biofilms similarly to the wild-type DegP activity, implying that protease-activity of DegP is not required to repress EHEC biofilms. In addition, biofilm related gene expressions (*e.g.*, curli, fimbriae) of EHEC cells were affected upon DegP interacting with OMPs on EHEC cells. This study elucidates a new extracellular function of DegP in controlling EHEC biofilms and provides us with an insight to develop an alternative strategy to control biofilm-related infections.

# 604697 Antifouling Topographies to Combat Biofilm-Associated Antibiotic Tolerance

**Abstract:**

**Antifouling topographies to combat biofilm-associated antibiotic tolerance**

Huan Gu[1,2], Sang Won Lee[1,2], and Dacheng Ren[1,2,3,4*]

[1]Department of Biomedical and Chemical Engineering, [2]Syracuse Biomaterials Institute, [3]Department of Civil and Environmental Engineering, [4]Department of Biology, Syracuse University, Syracuse, NY 13244, USA

Bacterial attachment to indwelling medical devices and the subsequent formation of multicellular biofilms, are the primary causes of hospital-acquired infections. Biofilms are composed of microbes encased in self-produced extracellular polymeric substrates (EPS). Due to the protection of EPS, a hallmark characteristic of mature biofilms is their high-level tolerance (up to 1,000 times) to conventional antibiotics compared to planktonic cells. Such multidrug tolerance makes it essentially impossible to eradicate mature biofilms, leading to persistent infections with high morbidity and mortality, and posing a heavy burden on the healthcare system.

To effectively control biofilms, it is important to understand the dynamic change in antibiotic susceptibility during biofilm formation and associated mechanisms. By tracking the susceptibility of *Escherichia coli* biofilm cells to antibiotics during the first 24 h biofilm formation, we found that biofilm cells were not always tolerant to antibiotics. Specifically, *E. coli* biofilms were formed in Lysogeny Broth (LB) on glass surfaces and treated with 200 μg/mL ampicillin (Amp) or 5 μg/mL ofloxacin (Ofx) for 1 h at 37 °C. The results show that the susceptibility of *E. coli* biofilm cells to both Amp and Ofx increased in the first 2.5 h after the inoculation. After this time point, the susceptibility to Amp decreased with the start of EPS production, while the susceptibility to Ofx did not decrease until 16 h after inoculation. Interestingly, the embedded biofilm cells remained active and thus susceptible to antibiotics if dispersed from the surface. Similar phenomena were observed during the biofilm formation of *Pseudomonas aeruginosa* PAO1 and Uropathogenic *E. coli* (UPEC) cells. These results indicate that physiological changes during early-stage biofilm formation can sensitize bacterial cells to conventional antibiotics. Thus, materials engineering to create dynamic and active topographies may render bacteria cells stay in an active stage and thus susceptible to antibiotics. We will present two examples of material design for such effects.

The first example is biofilm control using dynamic topographies generated by shape recovery of t-butyl acrylate (tBA) based shape memory polymer (SMP). This polymer can achieve one-way shape recovery from the temporary to permanent shape when the temperature increases from 25 to 40°C. We demonstrate that such on-demand actuation of shape recovery causes effective removal of both bacterial and fungal biofilms, e.g., 99.9% (3 logs) removal of *P. aeruginosa* biofilms. Along with biofilm removal, the detached cells were sensitized to antibiotics. For example, the susceptibility of *P. aeruginosa* to 50 μg/mL tobramycin increased by 2,479-times compared to the static flat control. This was attributed to both the disruption of biofilm structure and the increase in cellular activities as evidenced by the increase in intracellular level of ATP and expression of the *rrnB* gene.

Encouraged by the effects of dynamic topographies, we further developed a biofilm control strategy based on active topographies with magnetically driven micron-sized pillars. By turning the electromagnetic field on and off at a tunable frequency, active pillars can beat periodically, and thus, achieve long-term control of biofilm formation by inhibiting bacterial attachment and removing mature biofilms on demand (with stronger actuation). Active topographies with the optimized design prevented biofilm formation and removed established biofilms of UPEC, *P. aeruginosa*, and *Staphylococcus aureus*, with up to 3.7 logs of biomass reduction compared to the flat and static controls. In addition, the detached biofilm cells were sensitized to bactericidal antibiotics to the level comparable to exponential-phase planktonic cells. Such effects demonstrate a synergy between smart materials and antibiotics, which may help design safer medical devices.

# 611580 Computational Modeling and Experimental Characterization of Cathodic Voltage Controlled Electrical Stimulation (CVCES) for Eradication of Biofilms


**Abstract:**

Orthopedic implants used to replace or support a damaged bone or joint are susceptible to bacterial biofilm infections at the skin-contacting and bone-contacting sites, known as prosthetic joint infections (PJIs). These biofilms are hard to remove and are often resistant to conventional antibiotic treatments. This creates a need to design and develop new treatment procedures that can lead to complete eradication of biofilms on the surface of implants at a reduced financial cost and physical distress to patients. Electrochemical treatments are being developed for the treatment and prevention of orthopedic implant-associated infections. Cathodic Voltage-Controlled Electrical Stimulation (CVCES) is a minimally invasive technology for infection eradication without the need for implant removal. In CVCES, the metallic implant is utilized as a cathode and the biofilm is treated with a cathodic voltage. CVCES, in combination with antibiotics, has been shown to greatly reduce the bacterial burden on titanium implants in a rodent model of joint infection. Direct current flowing through the electrodes into the biofilm drives an electrolytic process that has a destructive effect on the bacterial biofilm. CVCES is a "voltage-controlled" process because the interfacial electrochemical properties of passivated metals utilized in orthopedics are voltage dependent. Because most bacteria have a negative surface charge, application of cathodic voltage leads to a negative charge-charge interaction resulting in electrochemical repulsion that promotes desorption of adherent bacteria from the electrode. This, along with the reduction reaction products, is why cathodic voltage is utilized to modulate the adjacent biological microenvironment.

Though CVCES is proven to be effective in biofilm eradication, many aspects and fundamental mechanisms with respect to the treatment protocol remain to be understood. In this work, we performed *in vitro* experiments and computational modelling to elucidate the key changes in the system leading to biofilm removal. An increase in pH and reduction in dissolved oxygen content near the cathode have been identified as two major factors that govern the electrochemical therapy. The mathematical models have been developed using the commercial multhiphysics software COMSOL (


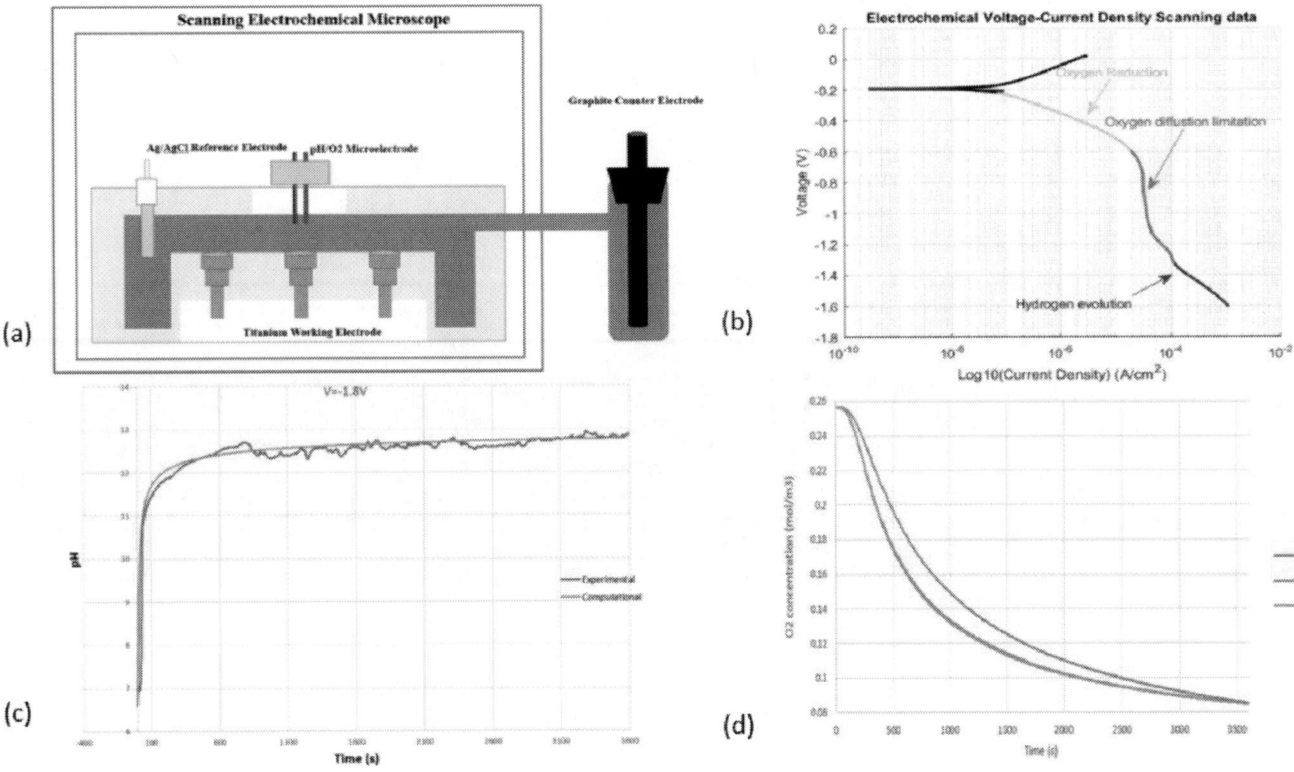


Figure 1: (a) Schematic of the *in-vitro* assay (b) cpTi polarization scan (c) pH profile distribution at 10 μm from the electrode at -1.8V (d) Computational $O_2$ concentration at 10 μm from the electrode

# 611360 Interfaceof Electrogenic Bacteria and Reduced Grapheneoxide: Energetics and Electron Transport

**Abstract:**

A synergistic, nanoscale electrical-interface with the membranes of exoelectrogenic microbes will have transformative impact on biological cell based electronic-devices. Here, we report that a conformal graphenic interface on biocatalytic Geobacter sulfurreducens membrane results in quantum-capacitance induced n-doping in reduced graphene oxide (rGO) that further enhances electron shuttling from the membrane to improve electron harvesting from the electrogenic organism. The quantum coupling of rGO with the connected protein-membrane channels leads to an additional electron density of 3.91 x $10^{12}cm^{-2}$ and an increase in the in-plane phonon vibration energies (G) of rGO by ~5 $cm^{-1}$. This n-doping enhances the electron transfer-rate from the cell membrane into the rGO via a net driving potential of 158 meV with a 3-fold increase in the power density.The synergistic electron-harvesting and conformal membrane-interfacing of flexible 2D nanomaterials can lead to an evolution in the design of microbe-circuitry to power stand-alone nanodevices.

# 614130 Finding a Niche – Microbial Colonization of Host Mucosal Barriers

**Abstract:**

Mucus is a biological gel that lines all wet epithelia in the body, including the mouth, lungs, and digestive tracts, and has evolved to protect us from pathogenic invasion. Microbial pathogenesis in these mucosal systems, however, is often studied in mucus- free environments, which lack the geometric constraints and microbial interactions that are found in natural, three- dimensional mucus gels. To bridge this gap, my laboratory has developed model test systems based on purified mucin polymers, the major gel-forming constituents of the mucus barrier. We use this model to understand how the mucus barrier influences bacterial virulence, and moreover, to elucidate strategies used by microbes to overcome the normal protective mucus barrier. I will discuss data showing that the mucus environment has a significant impact on the physiological behavior of microbes, including surface attachment, quorum sensing, the expression of virulence genes, and biofilm formation. The picture is emerging that mucins are key host players in the regulation of microbial virulence and can guide the fabrication of advanced polymers to regulate host-microbe interactions.

# 614315 Life in a Tight Spot: Transport and Collective Behavior of Bacteria in Heterogeneous Media

**Abstract:**

Diverse processes in healthcare, agriculture, and the environment rely on bacterial motility in porous media; indeed, most bacterial habitats—e.g. biological gels, tissues, soils, and sediments—are heterogeneous porous media. However, while bacterial motility is well-studied in homogeneous environments, how confinement in a porous environment impacts bacterial transport remains poorly understood. To address this gap in knowledge, we combine microscopy, materials fabrication, and microbiology to investigate how *E. coli* moves in 3D porous media. By probing single cells, we demonstrate that the paradigm of run-and-tumble motility is dramatically altered by pore-scale confinement. Instead, we find a new mode of motility in which cells are intermittently and transiently trapped as they navigate the pore space; analysis of these dynamics enables prediction of bacterial transport over large length and time scales. Further, by developing a new 3D printing approach, we design multi-cellular communities with precise control over the spatial distribution of bacteria. Using this approach, we show that concentrated populations can collectively migrate through a porous medium—despite being strongly confined—and develop principles to predict and direct this behavior.

# 614035 The Pseudomonas aeruginosa Biofilm matrix

**Abstract:**

A critical element of *Pseudomonas aeruginosa* pathogenesis is its ability to form biofilms in the lungs of CF patients. Biofilm bacteria produce one or more extracellular polymeric substances (EPS) that act as a scaffold, holding biofilm cells together and to a surface. We discovered that *P. aeruginosa* has the capacity to encode at least two alternative exopolysaccharides, designated Psl and Pel, which play a critical role in biofilm formation. Psl has been reported to consist of a neutral polymer of a pentasaccharide subunit. Recent discoveries by our group have determined that Pel is a positively charged amino sugar polymer. Biofilm cells are less susceptible to antimicrobials than their planktonic counterparts. While this phenomenon is multifactorial, the ability of the biofilm matrix to reduce antibiotic penetration into the biofilm is thought to be of limited importance, as previous studies suggest fairly rapid movement of antibiotics through biofilms. In this study, we monitored the transport of two clinically relevant antibiotics, tobramycin and ciprofloxacin, into non-mucoid *P. aeruginosa* biofilms. To our surprise, we showed that the positively charged antibiotic tobramycin is sequestered to the biofilm periphery, while the neutral antibiotic ciprofloxacin readily penetrated. We provide evidence that tobramycin in the biofilm periphery both stimulated a localized stress response and killed bacteria in these regions, but not in the underlying biofilm. Although it is unclear which matrix component binds tobramycin, its penetration was increased by the addition of cations in a dose-dependent manner, which led to increased biofilm death. These data suggest that ionic interactions of tobramycin with the biofilm matrix limit its penetration. We propose that tobramycin sequestration at the biofilm periphery is an important mechanism in protecting metabolically active cells that lie just below the zone of sequestration.

# 614316 Spatial Localization and Treatment of Individual Cells in Multispecies Biofilms

- Download/View Extended Abstract


**Abstract:**

Bacterial cells living at interfaces usually have neighbors of many different species and are able to organize themselves into complex structures named biofilms. A more effective management of these structures requires both an in-depth characterization of the spatial location of the individual cells and the use of molecules that are able to control their growth. In here, I will discuss the application of nucleic acid mimics (NAMs), also known as nucleic acid analogues, to study and manage these multispecies biofilms. More specifically, I will provide examples of works that are currently being carried out within our research team and that include: 1) coupling NAMs and fluorescence *in situ* hybridization for a more robust detection of pathogens within biofilms; 2) combination of NAMs and delivery vectors such as liposomes and cell-penetrating peptides for a more efficient treatment of individual microbial cells within biofilms.

# 614131 Biological Inspiration for Anti-Biofilm Strategies

**Abstract:**

In clinical environments, bacterial infections often take the form of interface-associated biofilms or otherwise aggregated communities. These communities exhibit cooperative behavior and emergent functions that confer on their constituent cells greater tolerance to antimicrobial treatments and lead to persistent infections. Methods to prevent or disrupt biofilms and bacterial aggregates, therefore, represent promising strategies to improve outcomes of conventional antimicrobial therapies and limit the spread of antibiotic resistance. Here, I will discuss molecular investigations of bacterial communication, competition, and resilience to stress involved in the establishment or maintenance of biofilms as biological inspirations for such strategies. Bacterial glycolipid surfactants induce selective permeability in bacterial membranes and exhibit cooperative antimicrobial activity with antibiotics. Iron competition between a gut pathogen and a probiotic bacterium in *in vitro* biofilm infection models is mediated by a molecular Trojan Horse strategy to combat piracy of secreted iron-chelating compounds. Finally, a ubiquitous biosynthetic pathway for osmotic stress response and energy storage in bacteria is critical to biofilm formation, and a small molecule inhibitor of a critical enzyme is effective in preventing biofilm development. These naturally occurring intra- and interspecies processes serve as biological inspiration for anti-biofilm and antimicrobial therapies.

# 613999 Biomimetic Drugs: *In Vitro, In Vivo* & mechanistic Studies of 5-12mer Antimicrobial & Antiviral Peptoids to Treat Respiratory Infections at the Airway Interface

**Abstract:**

Growing bacterial resistance to conventional antibiotics has spurred the exploration of bioengineered antimicrobial peptides (AMPs) and analogues as novel anti-infective agents. However, since peptide bioavailability is limited by proteolysis, non-natural AMP mimics are interesting as more robust and biostable antimicrobials, and offer distinct advantages as potential clinical therapeutics. We report here on our experimental exploration of the development of **poly-*N*-substituted glycines (peptoids)** as a new class of biomimetic antimicrobial drugs, via multiple approaches including several different *in vitro* assays and *in vivo* mouse studies. After studying more than 120 peptoid sequence variants, we identified a number of unique peptoids that exhibit potent, broad-spectrum antibacterial *in vitro* activity, and which have a unique, biomimetic mechanism of action, similar to that of the human antibacterial peptide LL-37: bacterial rigidification. In our most recent *in vivo* testing, mice were infected intratracheally with bioluminescent *Pseudomonas aeruginosa*, then treated by our TM5 peptoid, providing a significant reduction in bacterial loads compared to untreated animals. TM5 peptoid was also well tolerated in the lung by mice. In addition, new super-resolution fluorescence videomicroscopy studies confirm that these peptoids rapidly "rigidify" bacterial cytoplasm, just like natural cathelicidin AMPs (LL-37, cecropin). New experimental results show that like LL-37, as well, our peptoids have antiviral activity against HSV-1 as well. Finally, we have shown via SAXS studies that the activity and potency of cationic, amphipathic TM5 peptoids is enhanced by their self-assembly into stable, ellipsoidal micelles. **Taken together, these results show the highly promising potential clinical applicability of these 5-12mer peptoids.**

# 602740 Thinking Small for Pandemic Solutions: A Nanomedicine Platform for Diagnosing and Treating Viral Infections

**Abstract:**

At the time of writing this, the coronavirus COVID-19 pandemic is wreaking havoc on the world. Far too many times, the world is caught off-guard by a new virus without having accurate testing to determine who is infected and certainly without appropriate vaccines or other therapies to treat such viral infections. COVID-19 is just another example in a long list of recent viruses which includes Ebola, H1N1, etc. This presentation presents a paradigm shift in viral diagnostics and therapies moving our healthcare system from treating each virus as a separate microbe to treating them as a platform. Specifically, we introduce nanomedicine, or the use of nanomaterials, as a versatile platform that can be functionalized to attach to several active regions on a virus simultaneously (thus being able to deal more readily with virus mutations) and more effectively than today's single antibody or aptamer strategies. It will include results on novel nanoparticles, self-assembled materials, and viral theranostics that should be the focus of future medicine to avoid our constant "being caught off guard" healthcare system.

# AUTHOR INDEX


Agarwal, Harshit.....9
Albright, Victoria.....1
Aljaafari, Haydar.....13
Alkekhia, Dahlia.....5
Andersen, Marissa.....26
Andrianov, Alexander K......1
Aoki, Kazuhiro.....17
Arkin, Adam P......27
Azevedo, Nuno.....36
Bah, Aissatou.....25
Baker, William D......10
Balmuri, Sricharani.....18
Barron, Annelise.....7, 38
Bauer, Joschka.....11
Berry, Vikas.....32
Blackwell, Helen E......9
Blanco, Irene Andreu.....20
Bogan, Abner.....24
Borca-Tasciuc, Diana A......19
Bose, Arijit.....20
Chen, Wesley.....17
Cheng, Yifan.....19
Chicchelly, Hannah.....13
Chittur, Priya.....21
Clark, Caelen.....31
Cotts, Sheldon.....32
Datta, Sujit S......34
Delparastan, Peyman.....11
Deutschbauer, Adam M......27
Donadt, Trevor.....6, 14
Edson, Julius.....2
Ehrensberger, Mark.....31
Emrick, Todd.....22
Fang, Kuili.....28
Feng, Guoping.....19
Fink, Tanner D......11
Flores-Mireles, Ana L......26
Fong, Junie.....26
Garcia, Caterina Bartomeu.....3
Gu, Huan.....29
Hacker, Timothy A......9
Han, Wei.....23
Hochbaum, Allon I......37
Hong, Seok Hoon.....28
Howell, Caitlin.....26
Illouz, Sylvia.....27
Jones, A-Andrew.....16
Keisham, Bijentimala.....32
Kelkar, Shaunak.....3
Komatsu, Takafumi.....3
Kornfield, Julia A......21
Kurtz, Irene S......15, 22
Kwon, Young Jik.....2
Larose, Cassi.....5
Lee, Sang Won.....29
Libera, Matthew.....7
Lin, Jennifer S......7
Liu, Hanwei.....21
Lynn, David M......9
Machado, Mary C......20
Messersmith, Phillip.....11
Michalik, Patryck.....12
Moraru, Carmen I......19
Niepa, Tagbo H. R.....18, 25
Nuxoll, Eric.....13
Nyffeler, Kayleigh E......9
Oliveira, Tania.....20
Palecek, Sean P......9
Parnian, Parham.....13
Parsek, Matthew.....35
Phandanouvong-Lozano, Vienvilay.....18
Prince, Joshua.....16
Ren, Dacheng.....29
Ribbeck, Katharina.....17, 33
Ritchie, Stephen.....10
Saleh, Bahram.....3
Sathyan, Ashlin.....22
Scheibel, Thomas.....11
Schiffman, Jessica D......15, 22
Shukla, Anita.....5
Song, Fangchao.....27
Stewart, Elizabeth J......12
Sukhishvili, Svetlana A......1
Summers, Ryan M......10
Swihart, Mark T......31
Tiemeyer, Michael.....17
Tirrell, David.....21
Tripathi, Anubhav.....20
Usman, Huda.....25
Vishnoi, Priyanshu.....31
Waldman, Abraham J......15
Webster, Thomas J......3, 39
Werlang, Caroline.....17
Worobo, Randy W......19
Wu, Qiong.....23
Xiao, Xixi.....7

Yang, Rong........................................................................6, 14
Yeung, King Lun ..................................................................23
Yu, Chao.................................................................................5
Yusa, Fumie............................................................................3
Zha, Runye............................................................................11
Zhao, Wenhan.........................................................................7